The Mind-Stretching Adventures of Anna Lize and Saul Van Chek

Beyond Compare:
An Exploration of Ratios and Proportions

Written by Norm Lyons   Illustrated by Indigo Prasad

ISBN-13: 979-8-218-66866-2

Dedication

To all the folks out there who enjoy engaging with mathematics, rhyme, and wordplay in playful and creative ways...here's another one for you!

And, yet again, to Ollie and Ruby...long live dachshunds!

Finally, I, once again, deeply appreciate the wonderful and insightful editing of my transcript performed by my teaching colleague and friend, Dana Kirk.
Thanks so much, Dana!

N. L.

To all the young artists looking for their passion and finding ways to express themselves.

I. P.

"Off to the park!" says Anna Lize
To Saul Van Chek, "The game's today!
Let's cheer for our school friends who'll play
Their favorite sport under blue skies!"

1

These two best buddies leave Saul's place
To watch a baseball game at noon.
Their good friends' team will play real soon!
A tough opponent they will face!

As the kids reach the field they see
A large crowd sitting in the stands.
That's one huge throng of avid fans!
Saul says, "Let's go!" excitedly!

3

The two classmates go take their seats
With all of the spectators there.
The sun is shining, skies are fair.
A great day to watch these athletes!

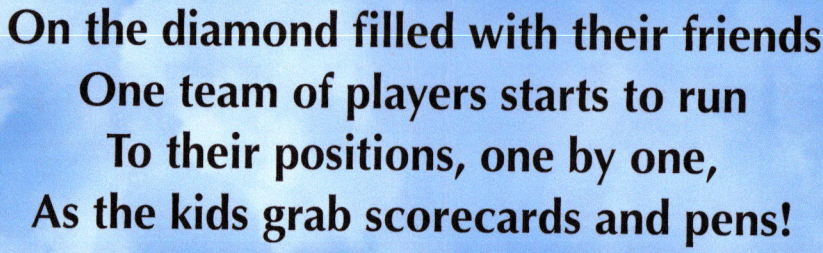

On the diamond filled with their friends
One team of players starts to run
To their positions, one by one,
As the kids grab scorecards and pens!

5

6

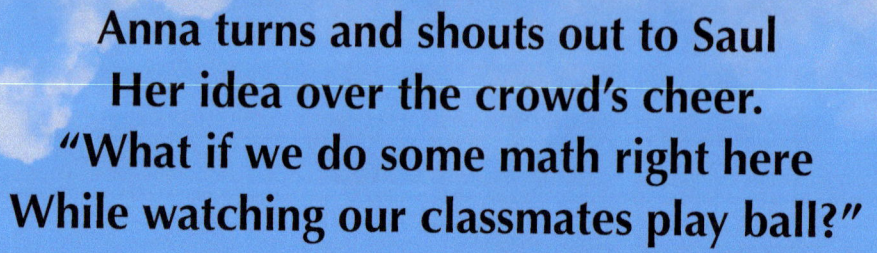

Anna turns and shouts out to Saul
Her idea over the crowd's cheer.
"What if we do some math right here
While watching our classmates play ball?"

Saul says, "Yes, that suggestion's great!
Math problem solving is my thing!
We'll 'talk math' while the batters swing
At baseballs headed for home plate!"

The blue team is the first to bat,
Up to the plate steps strong Pablo.
The pitcher gets ready to throw
And, bam! There's strike one, just like that!

8

At this point in time Anna states,
"Let's kick off our fun math journey
And only use the things we see
Here at this game with our classmates!"

So Anna says to Saul, her pal,
"The first topic we will explore
Is <u>ratios</u>! They're here, galore,
At this awesome ballpark locale!"

"Recall a ratio compares
The sizes of two quantities
With great efficiency and ease!
They're everywhere!" Anna declares.

"Remember ratios apply
To many, many different sorts
Of things, like recipes and sports,
And even to the things we buy!"

Saul states, "OK, let's find a few
Examples we can analyze.
As you've expressed it's no surprise
That there are lots at this venue!

"We'll first compare the number of
Kids playing on the field out there
Who have on their hand in the air
A black- or a tan-colored glove!

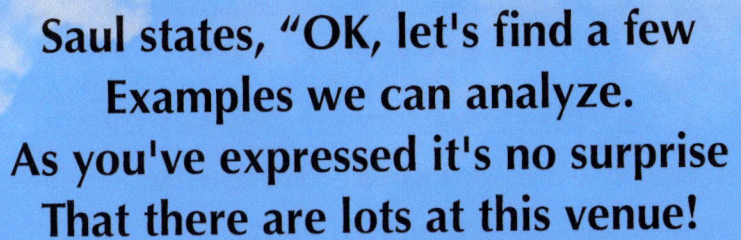

"In all there are six kids clearly
Sporting black gloves with such great pride.
The rest have tan gloves at their side,
Focused on fielding intently!

"The fitting ratio, therefore,
Of black to tan gloves, you can see,
Is none other than six to three
And, as you know, Anna, there's more!"

"Correct, Saul!" Anna states. "To start,
We'll use a fraction to express
Our ratio in this process,
Then simplify it 'cause that's smart!"

"Six over three's the fraction, no?
Or, simplified, two over one.
Hold on right there, we're not quite done!
Thus, two to one's the ratio!"

Black gloves to tan gloves
⬇
Black gloves : tan gloves
⬇
$6:3 \rightarrow \frac{6}{3} = \frac{2}{1} \rightarrow 2:1$

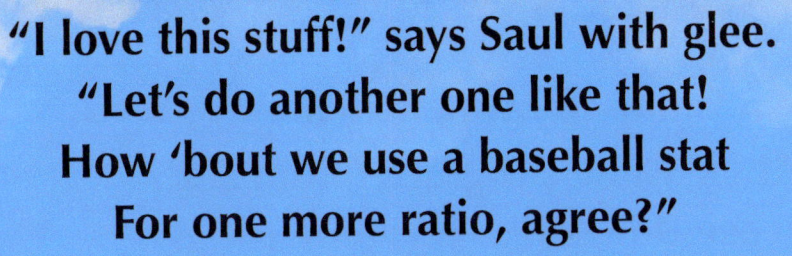

"I love this stuff!" says Saul with glee.
"Let's do another one like that!
How 'bout we use a baseball stat
For one more ratio, agree?"

"Of course!" replies Anna, "Let's talk
About the one you'd like to choose."
Saul says, "I know which one we'll use!
The ratio: strikeout-to-walk!"

14

So Saul and Anna take a look
At their game program full of stats
And see things like hits and at-bats,
And pitchers' info in this book!

15

Saul finds the data that relates
To Ernie Runn, who's on the mound,
And says, "This season Ernie's bound
To truly be one of the greats!

"Up to this day Ernie's amassed
Twenty-one strike-outs and three walks!
(He's also been charged with two balks!)
Watching him pitch is such a blast!

Ernie Runn

YR	Club	G	IP	W	L	ER	SO	BB	BK	ERA
24	Skylions	7	35	7	0	10	21	3	2	2.57

" We know that his strikeout-to-walk
Ratio's twenty-one to three,
Or seven to one more simply.
For his pitched games winning's a lock!"

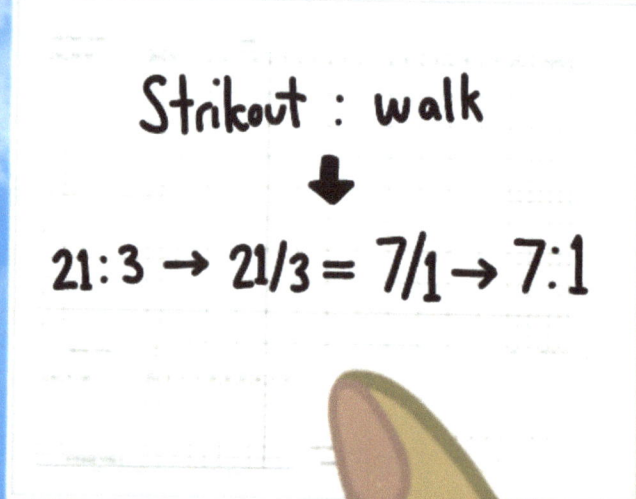

Anna remarks, "Let's next look at
Ratios in an equation,
Raising our math conversation,
Moving beyond the baseball stat!"

Saul states, "Sounds good! First, we'll review
Proportions which, formulated,
Are two ratios equated
And which are used in life a slew!"

18

"Consider the concession stand
And note two hotdogs have a price
Of four dollars, to be precise,
With loads of free toppings on hand!

"If I buy hotdogs at that spot
And pay exactly twelve dollars,
My key question to us scholars
Is, 'How many hotdogs were bought?'"

1 hotdog → $2
2 hotdogs → $4

20

Anna's eager to solve this now!
She shares her shrewd approach with Saul.
"We'll use a proportion, that's all!"
Saul says, "Of course! Please explain how!"

"We'll make two ratios," she states,
"And compare price to quantity
Of hotdog purchases, you see.
The final answer still awaits!

"Those ratios we'll equate, too.
Our model is set in cement!
Four over two's equivalent
To twelve over the 'x' value!

"That 'x' value? It represents
The unknown hotdog purchase count
For the twelve-buck payment amount!
I hope, Saul, that this all makes sense!"

$$4/2 = 12/x$$

$x =$ # hotdogs purchased for $12

"Of course it does, Anna! Stand back!
Don't forget math's *my* expertise!
I'll solve this problem with great ease
And use a smart plan of attack!

$$4/2 = 12/x$$

$x = \#$ hotdogs purchased for \$12

"At this point I'll cross-multiply
To form a new equation next:
Two times twelve equals four times 'x,'
Which yields 'x' equals six! See why?"

$$\frac{4}{2} = \frac{12}{x}$$

$$\frac{4}{2} \times \frac{12}{x} \rightarrow 2 \cdot 12 = 4 \cdot x$$

$$24 = 4x \rightarrow x = 6 \text{ (hotdogs)}$$

Anna asserts, "Saul, way to go!
Your math's truly beyond compare!
'Six hotdogs' was the answer there!
Your solution was quite thorough!"

"Thanks, Anna!" Saul says with a smile.
"Let's solve one more proportion here!
That drink stand at the ballpark's rear
Is making lemonade with style!

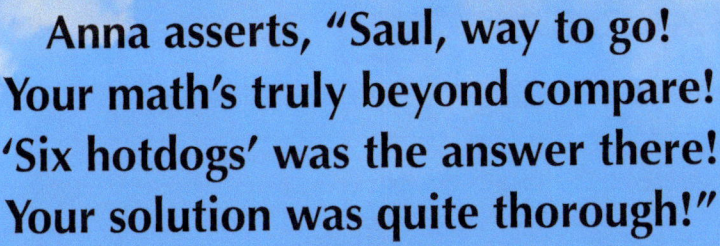

"I've closely watched their recipe:
Three-quarters cup of lemon juice
For two quarts of drink they produce.
It's making me very thirsty!

28

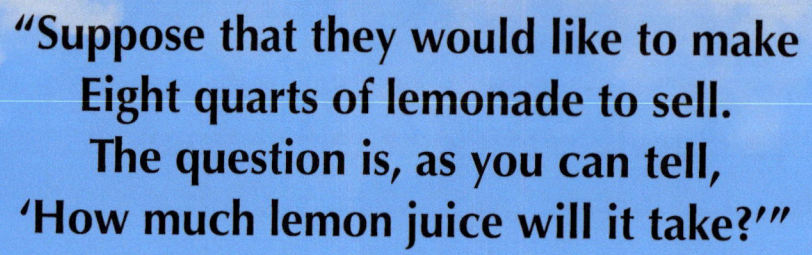

"Suppose that they would like to make
Eight quarts of lemonade to sell.
The question is, as you can tell,
'How much lemon juice will it take?'"

"OK," says Anna, "let's reprise
Setting up ratios to solve
A proportion that will involve
Comparing liquid quantities!

"Three-quarters over two's the first
Ratio that we will equate.
The second one's 'x' over eight!
Want lemonade to quench your thirst?"

$$\frac{3/4}{2} \qquad \frac{x}{8}$$

x = # cups of lemon juice for 8 quarts of lemonade

Saul replies, "Yes!" and starts to run
With Anna towards the small drink stand
To buy some fresh lemonade and
Finish solving the proportion!

32

Lemonade

1 cup → $1.50

33

With a cup brimming Anna thinks,
And next Saul hears his friend exclaim,
"Three-quarters times eight is the same
As 'x' times two!" Then, Anna drinks!

She continues, "The 'x' value
Matches with three cups lemon juice!
Math makes things easy to deduce
And always helps you work things through!"

$$\frac{3/4}{2} \diagdown \frac{x}{8} \rightarrow 3/4 \cdot 8 = x \cdot 2$$
$$6 = 2x \rightarrow x = 3$$

(cups of lemon juice)

34

Each kid heads back up to their seat
To watch the end of the ballgame.
A base hit won it! What a shame
For the team that suffered defeat!

Anna and Saul leave the ballpark
By passing through the exit gate.
That's where Anna's mom and dogs wait
To walk the kids home before dark!

36

At Saul's place when the day is done,
Anna and Saul say their good-byes.
These two feel they're a bit more wise
'Cause today was a math homerun!

37

38

The End